**The Impact of
Science and Technology**

SPACE EXPLORATION

Joseph Harris

FRANKLIN WATTS
LONDON•SYDNEY

First published in 2009 by Franklin Watts

Copyright © 2009 Arcturus Publishing Limited

Franklin Watts
338 Euston Road
London NW1 3BH

Franklin Watts Australia
Level 17/207 Kent Street
Sydney, NSW 2000

Produced by Arcturus Publishing Limited
26/27 Bickels Yard
151-153 Bermondsey Street
London SE1 3HA

Series concept: Alex Woolf
Editor and picture researcher: Nicola Barber
Cover design & illustration: Phipps Design
Consultant: Brian Clegg

A CIP catalogue record for this book is available
from the British Library.

Dewey Decimal Classification Number: 629.4' 1

ISBN 978 0 7496 9223 0

Printed in China

Franklin Watts is a division of Hachette
Children's Books, an Hachette UK company.
www.hachette.co.uk

Picture credits
Corbis: 6 (Stefano Bianchetti), 10, 30 (Charles O'Rear),
45 (Sergei Remezov/Reuters), 46 (Gene Blevins), 51 (Xinhua).
Science Photo Library: cover (NASA/JPL-Caltech/University of
Arizona), 9 (Detlev van Ravenswaay), 13, 53, 59 (Victor Habbick
Visions), 14, 20, 24, 48 (NASA), 19 (Roger Harris),
23 (Chris Butler), 25, 26 (NASA/JPL/UA/Lockheed Martin),
32 (Cheryl Power), 37 (European Space Agency),
43 (CNES, 2004 Distribution Spot Image), 55 (Dr Seth Shostak),
57 (David A. Hardy).
Shutterstock: 5 (Giovanni Benintende), 17 (Jennifer Scheer),
28 (Artifan), 34 (Michelle D. Milliman), 38 (Daniel Padavona),
41 (Vladislav Gurfinkel).

Cover picture: the *Phoenix* spacecraft lands on Mars on
25 May 2008.

Every attempt has been made to clear copyright. Should there
be any inadvertent omission, please apply to the publisher for
rectification.

CONTENTS

CHAPTER 1
The Science of Space   4

CHAPTER 2
The Politics of Space   8

CHAPTER 3
Living in Space   16

CHAPTER 4
Unmanned Exploration   22

CHAPTER 5
Space Technologies and
Everyday Life   28

CHAPTER 6
Satellites   36

CHAPTER 7
The Economics of Space   44

CHAPTER 8
Our Future in Space   52

Glossary   60
Further Information   63
Index   64

The Science of Space

Our universe is vast. It contains billions of galaxies, and these in turn contain hundreds of billions of stars. Some of the stars are orbited by planets. Despite the presence of so much matter, most of the universe is made up of space. Space is the void or emptiness that contains and separates planets, stars and other bodies. Humans have explored Earth and its atmosphere, but we have travelled hardly any distance into the rest of space. That is why space is often described as the 'final frontier', a phrase famously used in the science-fiction TV show *Star Trek*.

The final frontier

Exploring space offers unique challenges and rewards. Space is a vast and unwelcoming vacuum. A perfect vacuum would be absolutely empty, with no matter at all, but space does in fact contain some gas molecules. Nevertheless, space is as close to a perfect vacuum as exists. It is silent, because sound does not travel through a vacuum. No unprotected human being can survive in space for more than a minute or two.

The solar system

The sun is one of billions of stars. Our solar system is the region of space in which every material body orbits the sun. Our Earth is one of these bodies. According to current definitions, the solar system contains eight planets: Mercury, Venus, Earth, Mars, Jupiter, Saturn, Uranus and Neptune. The planets are not the only bodies orbiting the sun. There are also dwarf planets, one of which, Pluto, was classed as a planet until 2006. There is also a large asteroid belt (a region of rocks smaller than planet size) which lies between Mars and Jupiter.

The distances between stars and galaxies are so great that they are measured in light years. A light year is the distance that light – the fastest thing in the universe – can travel in one year: 9.46 trillion kilometres. Even our sun's closest star, Proxima Centauri, is around 4.22 light years away. The Andromeda Galaxy is around 2.5 million light years away. This gives some idea of the mind-boggling size of the universe. In order to travel these distances, astronauts would need spacecraft capable of moving unimaginably faster than anything in service today.

Such huge distances mean that scientists on Earth study most of the universe from far, far away. When astronomers observe Proxima Centauri, the light has taken 4.22 years to reach the lenses of their telescopes. They are looking at the star as it was four years ago. When people look at pictures of the Andromeda Galaxy, they are looking millions of years back in time.

The Andromeda Galaxy is one of the closest galaxies to our own, yet it is still 2.5 million light years away. When we look at it through a telescope, we are seeing it as it was millions of years ago.

This print shows the 'Copernican system' as it was understood in the seventeenth century. The sun is at the centre of the solar system, orbited by the Earth and the other planets known at that time.

Studying the universe

Even before telescopes existed, people looked out into space and tried to understand the universe. The Egyptians, Babylonians, Greeks and Maya were among the first peoples to map the movements of the stars. In Europe, a sixteenth-century Polish mathematician called Nicolaus Copernicus put forward a revolutionary idea. Copernicus watched the stars and took measurements by eye. He concluded that the view of the universe accepted at the time – that the sun travelled round the Earth – was wrong, and that it was in fact the Earth that travelled round the sun. At the end of his life, he published this theory in a book.

Later astronomers built on Copernicus' work. Technological improvements led to the development of the telescope in the early seventeenth century. Using a telescope, the Italian scientist Galileo Galilei discovered the four largest moons of Jupiter and observed craters on the Earth's moon. He also confirmed that Copernicus' theory was correct. Over the centuries, scientific discoveries and information from telescopes advanced our knowledge of how the universe works. Telescopes became incredibly powerful, and they are still vital tools for studying the universe.

⊕ PROS: SCIENTIFIC REVOLUTION

Copernicus, and the astronomers and scientists that followed, revolutionized our understanding of the universe. Their work led to modern space sciences and space exploration. Scientists have found evidence of the Big Bang, a massive explosion that created the universe. They have established that the universe is expanding outwards, and have uncovered new astronomical phenomena such as quasars (extremely distant starlike objects) and pulsars (rotating neutron stars that emit pulses of radiation). Scientists have also refined their estimates of the age, size and shape of our universe. Thanks to the scientific efforts of past centuries, it is possible today to set off into the final frontier.

⊖ CONS: SCIENTIFIC REVOLUTION

People often find new ideas threatening, especially when their old ones are backed up by tradition. The Catholic Church of Galileo's day was horrified by his theories. It forbade Catholics to read Copernicus' book, because the theory that the Earth orbited the sun seemed to contradict the teachings of the Bible. Today, tensions can still arise between religion and science. For example, some groups do not accept the Big Bang explanation of the creation of the universe, because the Bible states that God created everything in seven days.

The importance of space exploration

VIEWPOINT

According to the renowned British physicist Stephen Hawking, part of what makes our species unique is humanity's drive to explore and learn about the universe:

'We are just an advanced breed of monkeys on a minor planet of a very average star. But we can understand the universe. That makes us something very special.'

CHAPTER 2

The Politics of Space

Exploring the final frontier has a political aspect. Nations have competed for centuries to control land, sea and – since the twentieth century – air. Now the question is, who will successfully exploit space to their advantage? Will all of humankind share the benefits of space?

The space race

The first steps into space took place in the 1950s. That decade was overshadowed by the Cold War, an intense competition for superiority and influence between the superpowers of the time. These rival powers were the United States and the USSR or Soviet Union (a gigantic communist state, centred on present-day Russia, which collapsed in 1991). Both sides feared that the other might gain an advantage which might prove crucial if war broke out. So the development of space exploration, by one side or the other, had huge political significance.

In 1957, the USSR surprised the world by carrying out the first successful spaceflight. The unmanned Soviet craft, *Sputnik 1*, made history by becoming the first human-made object to escape Earth's gravity and achieve orbit. *Sputnik* was visible from the ground as a tiny glimmering dot. Some people in the West viewed the satellite's passage across the sky with unease. They already lived in fear

Rockets

Rockets, the powerful engines used to propel spacecraft, were first developed as weapons of war. Towards the end of World War II (1939–45), scientists in Nazi Germany devised the V-2 rocket, which inflicted considerable damage on Britain. It was the first rocket capable of flying high enough to reach space. After the war, the United States and USSR tried to recover as many German V-2s as possible, and both countries used the technology as the basis for their own rockets. With the help of a German scientist, Werner von Braun, who was one of the original designers of the V-2, NASA developed the *Saturn V* rocket that carried *Apollo 11* to the Moon.

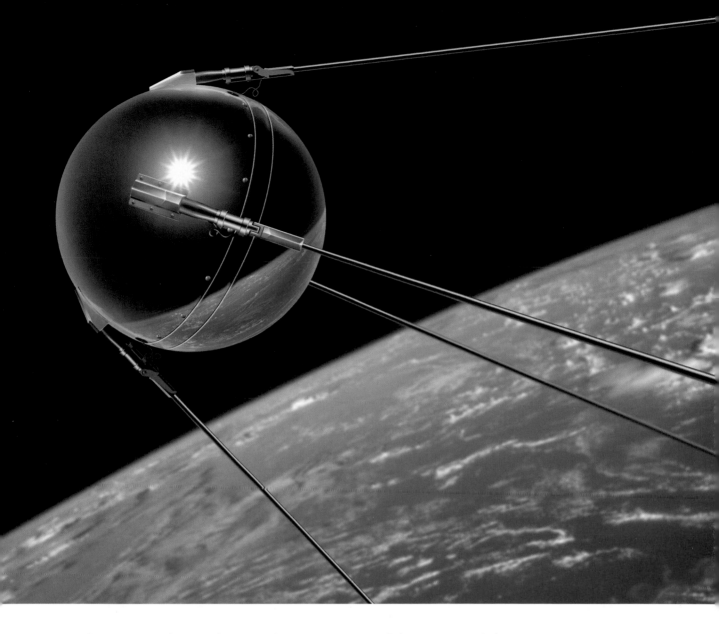

of Soviet missiles, so the USSR's new space capability suggested the possibility of attack on a new front. In response, the Americans redoubled their efforts to reach space. They formed a new organization, the National Aeronautics and Space Administration (NASA), in 1958. The superpowers began a 'space race' for superiority on the new frontier.

Both sides achieved successes. NASA launched its own unmanned craft, *Explorer 1*, less than four months after *Sputnik*'s flight. However, the Soviets maintained their lead, sending the first man into space in 1961. Yuri Gagarin orbited Earth in his craft, *Vostok 1*, and returned safely. In 1963, the USSR achieved another record when Valentina Tereshkova became the first female astronaut to travel in space.

Sputnik 1, **the first artificial satellite to orbit the Earth, transmitted radio signals back to Earth for 21 days.**

The Soviets' successes in space led the US president John F. Kennedy to set his nation's sights on a new goal. In 1961, he announced his determination to place a man on the Moon by the end of the decade. This NASA initiative was named the Apollo programme. On 20 July 1969, after many preliminary missions, *Apollo 11* landed on the Moon. NASA astronaut Neil Armstrong became the first human to set foot on the lunar surface. He declared it a "giant leap for mankind".

On 20 July 1969, NASA astronaut Edwin 'Buzz' Aldrin saluted the American flag as he stood on the surface of the Moon.

➕ PROS: SPACE COMPETITION

Armstrong's 'giant leap' owed much to politics. Competition between the superpowers created the space race and drove advances and achievements. The United States' attempts to blast an object into space had all failed before *Sputnik* was launched. After the launch, the United States was spurred on by an urgent desire not to be outdone. Political competition invigorated the space programmes of both powers.

The space race reflected the hostility and mistrust of the Cold War period. The West reacted with alarm to the *Sputnik* launch. This suggests that people at the time saw space competition as part of the superpower conflict on Earth, and spacecraft as a new threat.

The importance of the space race

VIEWPOINT

In 1967, reflecting on America's ambitious space programme, President Lyndon B. Johnson acknowledged the benefits it had brought to his country:

'We've spent between US$35–40 billion on space ... but if nothing else had come from that program except the knowledge that we get from our satellite photography, it would be worth ten times to us what the whole program has cost. Because tonight I know how many missiles the enemy has and ... our guesses were way off. And we were doing things that we didn't need to do. We were building things that we didn't need to build. We were harboring fears that we didn't need to have.'

Weapons in space

Once people were able to put objects into orbit, using space to make war became a possibility. In the movies, spaceships go into battle, firing lasers at one another. This is still a fantasy. But in the 1980s it seemed likely that deadly weapons could be sent into orbit to point at targets on Earth. In 1983, the US president Ronald Reagan announced plans for a defence system called the Strategic Defence Initiative (SDI). It was often jokingly called 'Star Wars', after the well-known series of space-adventure films. If the Soviets attacked the United States, the SDI would use a combination of Earth- and space-based technologies, including laser beams, to destroy enemy missiles in flight. Despite the president's determination to develop this system, the plans never went into effect.

American interest in space-based missile defences revived under the presidency of George W. Bush (2001–09). Although most of the focus was on defensive systems, there were some more aggressive ideas, such as the so-called 'Rod from God' satellite. This orbiting weapon would drop a metal cylinder from space on to a target on Earth. By the time the cylinder reached the ground it would be travelling at an enormous speed, and would annihilate its target.

Two other present-day powers, Russia and China, have the know-how to wage war in space. China has already shown that it can shoot down satellites from space. It is the third nation (after the United States and Russia) to have this ability. Several other countries are capable of launching satellites, and could also some day launch weapons into orbit.

➕ PROS: WEAPONS IN SPACE

Space-based weapons could improve global security. Most plans have followed the SDI model, and have been intended mainly for defensive purposes. The ability to destroy enemy missiles before they reach their target is a very attractive prospect. A space-based missile shield might provide the answer, particularly given the limited reliability of Earth-bound anti-missile defences.

➖ CONS: WEAPONS IN SPACE

Malfunctions and technical problems could have serious consequences on Earth. An accident or misunderstanding might even start a nuclear war. It is also difficult to guarantee that an aggressive regime would not turn its supposedly defensive system to offensive use. Another fear is that once one nation has developed space weaponry, others will feel they must do the same. Even though the Cold War is over, the Russians oppose the idea of the United States putting weapons in space, and have threatened to match any such move with weapons of their own. China might follow close behind. A situation like this could increase the chances of war on Earth, and obstruct the peaceful exploration of space.

The dangers of weaponizing space

Fears over space weapons focus on the possible consequences of starting a new arms race in space:

'There are no commanding heights in space which, once seized, can assure enduring advantage or dominance or which might prevent an arms race in space. None. The United States is not safer today for having initiated the development of the ultimate weapon [nuclear weapons] 60 years ago and for having sought to maintain sole possession of these weapons. To the contrary, the American government and the American people worry day and night that these weapons may come back at them and at others. This same fate will in time attend the possible weaponization of space. That is why it is a vital interest of all governments and all peoples to prevent that weaponization.'

(Jonathan Dean, retired US ambassador)

War in space. This computer-generated picture shows two armed satellites firing high-energy laser beams at targets on the Earth's surface.

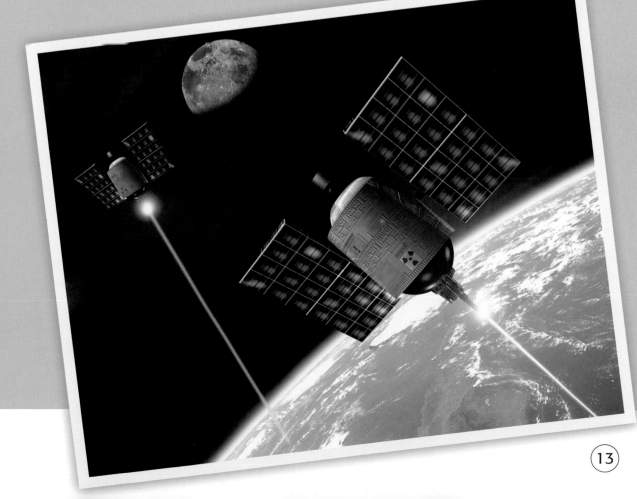

International cooperation in space

The exploration of space has provided some striking examples of international cooperation. Even during the Cold War, space projects provided opportunities for peaceful collaboration, and offered a way to improve relations between the rival superpowers. In 1975, the USSR and the United States undertook the joint *Apollo-Soyuz* mission. The spacecraft of both nations cooperated in various scientific efforts, such as photographing the sun. The mission was part of an attempt to ease the tensions of the Cold War (a process known as détente).

More recently, international cooperation has continued to be a feature of space exploration. During the 1990s, Russian and American astronauts shared experiences. Russians travelled on American space shuttles and Americans visited the Russian space station, *Mir*. These operations paved

US astronaut Peggy Whitson and Commander Valery Korzun from the Russian Space Agency enjoy a meal aboard the International Space Station in 2002.

the way for work on the current International Space Station (ISS), an ambitious project which has been under construction since 1998. Eighteen nations are involved in creating the ISS, which in 2001 received an award for being an outstanding example of international cooperation.

The benefits of international cooperation

VIEWPOINT

In remarks made at NASA headquarters in 2004, US president George W. Bush expressed his commitment to fostering international cooperation in space efforts:

'We'll invite other nations to share the challenges and opportunities of this new era of discovery. The vision I outline today is a journey, not a race, and I call on other nations to join us on this journey, in a spirit of cooperation and friendship.'

➕ PROS: INTERNATIONAL COOPERATION

Cooperative missions in space encourage understanding and communication between nations on Earth. A strong tradition of international cooperation in space could head off future conflicts. Space may become increasingly valuable, as new technologies make it possible to tap its natural resources. If spacefaring powers are willing to cooperate, all humanity is likely to benefit from the knowledge and resources gained from space exploration.

➖ CONS: INTERNATIONAL COOPERATION

Some people fear that sharing scientific knowledge with other nations could backfire. If the situation changed and the nations came into conflict, one side might be able to use inside information about the other's technical capabilities to its advantage.

CHAPTER 3

Living in Space

In 1961, Yuri Gagarin made his brief trip into space. His flight proved that, with the right technology, humans could survive in space and return safely to Earth. Since then astronauts have spent increasingly longer periods in space. Space agencies have built orbiting structures called space stations – special outposts, designed so that astronauts can live in them for a long time. The Russian astronaut Valeriy Polyakov holds the current record for the longest stay in space, having spent nearly 438 days aboard the space station *Mir* from 1994 to 1995.

The space shuttle

In the early 1980s, NASA introduced a new type of spacecraft, the space shuttle. It was much larger than previous vessels, providing more space for both crew and cargo. Unlike Gagarin's *Vostok 1*, which had only a cramped compartment for its single occupant, the space shuttle could comfortably house a crew of seven. Because of these features, the space shuttle was suitable for long missions in space.

NASA developed the space shuttle as part of an effort to design a reusable space vessel. In standard launches, rockets such as the American *Saturn V* and the Russian *Proton* carried a small crewed module – an independently-operated unit of the spacecraft.

Spacesuits

Astronauts occasionally need to go outside their craft into space. An operation of this kind is known as Extra-Vehicular Activity (EVA). Because space is a vacuum, astronauts can survive only by wearing a special suit. This spacesuit is incredibly tough and well insulated to protect the wearer from extreme temperatures. It covers the astronaut's entire body, with a transparent helmet over the head. The helmet is airtight, and an oxygen supply is pumped from canisters that the astronaut wears in a backpack. Also in the backpack is a water supply connected to the astronaut's mouth by a tube running into the helmet.

Once in orbit, the rockets fell away from the module and were lost for good. This was an extremely expensive and wasteful operation.

A reusable spacecraft can return intact to carry out many more missions. The space shuttle goes a long way towards this ideal. Its external fuel tank, which contains the liquid hydrogen fuel required to propel the shuttle into space, is destroyed during the launch. But the space shuttle's main component, a large module called an orbiter, is designed to survive the journey into space and back. Since 1981, orbiters have touched down in Florida and California more than 100 times. In 2004, it was announced that the space shuttle would be retired in 2010 and replaced by a new generation of spacecraft soon afterwards.

The orbiter of the space shuttle *Discovery* strapped to a *Saturn V* rocket and ready for launch. *Discovery* has been flying missions since 1984, and is the oldest shuttle still in use.

➕ PROS: SPACE SHUTTLE

The space shuttle provides a more cost-effective means to go into space than earlier vehicles. It can carry a larger crew and its size makes it versatile. It has an 18-m cargo bay (its cargo is often called its payload), and this can be used to carry satellites, space station components or scientific equipment. The space shuttle's payload bay can itself be used as a laboratory in which to conduct experiments in space. The crew can perform complex operations in orbit using equipment on the shuttle such as its mechanical arm.

➖ CONS: SPACE SHUTTLE

NASA abandoned its early plans for a fully reusable spacecraft because the development costs were too high. The shuttle was less expensive to develop, but is more expensive to run. Critics believe that greater investment should have been made to produce a more economical and state-of-the-art craft. The shuttle also has an imperfect safety record. Both the *Challenger* (1986) and *Columbia* (2003) disasters involved the tragic loss of their crews.

Space stations

Astronauts cannot stay in space for very long periods on spacecraft. The limited room restricts the amount of food, water and oxygen these craft can carry. Orbiting space stations are designed to provide a more spacious and long-term living environment. At present, space stations represent the nearest thing to permanent centres of human habitation in space.

The Soviet Union deployed the first basic space station, *Salyut*, in 1971. But

Modular space stations

Instead of being launched in a single, fully-assembled piece, space stations are constructed out of multiple sections, called modules. These modular stations are carried into space in stages, one unit at a time, and the units can be very large. The Russians pioneered this technique to build the space station *Mir*, and the same approach has been adopted for the current International Space Station since its launch in 1998. The modular design is perfectly suited for a cooperative project, since different nations can contribute separate modules.

it was in 1973, during the American *Skylab* mission, that astronauts first tested the effects of living in space for an extended period. Since then, astronauts have studied the effects of weightlessness on the human body and carried out scientific experiments in weightless conditions (more correctly called microgravity). Such experiments were performed aboard the Russian station *Mir*, which was launched in 1986 and remained in orbit until 2001. Today, investigations continue on the International Space Station.

The International Space Station: a computer-generated view of what the station will look like when the final modules have been put in place.

Microgravity

Space station missions have provided a huge amount of information about the way that microgravity affects the human body. Human beings evolved on Earth, so their biological processes are designed to function in Earth-like conditions. Gravity is one of the constants of life on our planet. This means that the weightless living conditions on board a space station have a range of effects on humans, some more serious than others.

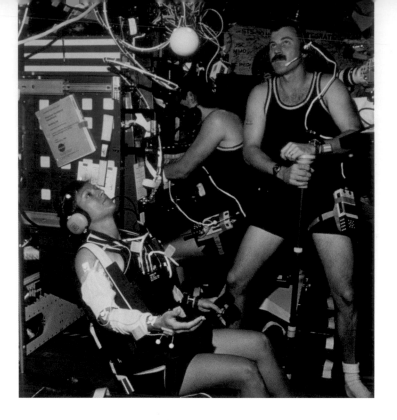

The most immediate challenge for astronauts living in orbit for long periods is to prevent their muscles from wasting away. This happens because muscles do not have to work so hard in an environment where everything is weightless. Astronauts must follow a strict programme of daily exercises to counteract this effect. Some of the other recorded effects of microgravity are more mysterious and more difficult to deal with. In space, the number of red blood cells, which carry oxygen from the lungs to the rest of the body, falls. Bones become less dense, which affects the strength of the skeleton. The immune system, which enables the body to fight diseases, is also weakened. Astronauts may also suffer from sleep disorders and may feel disoriented or mentally confused.

Astronaut Kathryn Hire is about to catch a ball during an experiment in the space shuttle *Columbia*'s Neurolab facility.

Is the ISS worth it?

VIEWPOINT

Not everyone agrees that the International Space Station plays a useful role. Some people believe that earlier space station missions have already revealed enough about the effects of microgravity on the human body. They argue that attention should focus on travelling beyond Earth's gravity:

'The only problem with this US$156 billion manifestation of human genius – a project as large as a football field that has been called the single most expensive thing ever built – is that it's still going nowhere at a very high rate of speed. And as a scientific research platform, it still has virtually no purpose and is accomplishing nothing.'

(Michael Benson, US space science commentator)

+ PROS: SPACE STATION RESEARCH

The effects of spending a long time in space are similar to those of natural ageing on Earth. Research into these effects has produced helpful results. For example, studies of astronauts' reduced bone density have improved our understanding of brittle bones (osteoporosis), which mostly affect older people. Treatment now emphasizes the importance of activity and impact exercise to maintain bone density.

Microgravity conditions can be used to manufacture materials in space. Substances of different densities, which would separate on Earth due to gravity, can be mixed in space. Furthermore, materials can be combined floating free without being in contact with a container. This helps the production of certain metals and glasses that have to be kept very pure. Microgravity conditions are also useful for biotechnology (the artificial creation and manipulation of living tissues). Biotech experts can grow proteins that cannot be produced on Earth. Space manufacturing is likely to be an exciting future growth area.

– CONS: SPACE STATION RESEARCH

A space station is not a natural environment for human habitation. It takes several years for an astronaut's bone density to return to normal after a long stay in space. Astronauts also experience psychological pressure in the cramped and strange environment. It is quite normal for people to feel very unwell before they acclimatize to being in space. Astronauts measure the severity of their symptoms on the 'Garn scale', jokingly named for US senator Jake Garn, who suffered the worst recorded case during his trip in 1985. Some people think that adapting to these conditions is a worthwhile challenge. Others question whether it is a good idea to put people into such a hostile environment. Space stations are also criticized for their cost and environmental impact. Spacecraft have to be launched frequently to restock the stations, and few of the materials on board can be recycled.

CHAPTER 4

Unmanned Exploration

Some scientists explore space without ever leaving their laboratories. They use powerful telescopes on the ground or in orbit to observe other stars and galaxies. Mechanical probes have been launched to gather data and send pictures of other planets back to Earth. Some probes transport robots that land on the surface of a planet. These activities supplement manned missions and offer an alternative way to explore space.

Telescopes

Telescopes were the first tools used for space exploration, and they remain important today. There are many different types of telescope. The most straightforward, the refractor, uses only a lens to gather light. However, there is a limit to how big a lens can be, and how far it can see. Today, refractor telescopes are primarily used by amateur astronomers.

Reflector telescopes use mirrors to gather light. Large reflector telescopes, such as the twin telescopes of the Keck Observatory in Hawaii, can gather enormous amounts of light. The mirrors at Keck are 10m in diameter, and each one is separated into 36 segments. The mirrors are divided in this way to prevent them cracking under their own weight.

Some modern telescopes do not observe visible light. The light we see is just one small part of the electromagnetic spectrum. The other parts (called wavelengths) consist of other forms of energy: gamma rays, X-rays, ultraviolet light, infrared

The Hubble Space Telescope

In 1990, the Hubble Space Telescope, a joint project involving the European Space Agency (ESA) and NASA, was launched into orbit. Because it is outside the Earth's atmosphere, Hubble has taken extremely clear images in the visible light portion of the electromagnetic spectrum, including striking images of faraway galaxies. Data from Hubble has allowed physicists to test their theories and expand their knowledge. For example, they have used images from Hubble to calculate the rate at which the universe is expanding.

light, microwaves and radio waves. Although human eyes cannot see these forms of energy, special telescopes can. Different astronomical phenomena, from quasars to clouds of hot gas, are most easily detected at certain e l e c t r o m a g n e t i c wavelengths. This means that gamma ray, X-ray, ultraviolet, infrared, microwave and radio telescopes, as well as optical telescopes, are all invaluable for learning about the universe. As many of these wavelengths are blocked or muffled by Earth's atmosphere, they are best observed from very high altitudes or, better still, from space.

The Hubble Space Telescope pictured in front of the spiral galaxy M100. The Hubble telescope is expected to be in service until about 2013, at which point the James Webb Space Telescope is scheduled to replace it.

➕ PROS: TELESCOPES IN ORBIT

Space-based telescopes gather data that Earth-based telescopes cannot. The Hubble telescope has produced stunning pictures of distant stars and greatly expanded our knowledge of the universe. It has even helped scientists calculate the age of the universe.

➖ CONS: TELESCOPES IN ORBIT

Space telescopes are difficult to repair. Hubble was launched in 1990 but the pictures it took were out of focus until 1993, when astronauts were able to fix the problem by replacing an incorrectly-shaped mirror. Other space shuttle missions to make repairs to Hubble have followed, but these are always costly and dangerous.

Probes

While telescopes have provided data on faraway stars and galaxies, space probes have gathered information about Earth's solar system. Probes are like other spacecraft, but with one crucial difference: there are no people in them. These robotic explorers are designed to operate autonomously (on their own) in space. There are cameras and scientific instruments on board, allowing them to record information about other worlds. Some probes travel to their destination planet and observe it from orbit. Others land on the planet or drop off a second probe, called a lander, which goes down to explore the surface.

There have been many probes since the Soviets carried out the groundbreaking *Luna 1* mission to the Moon in 1959. Probes have targeted different celestial bodies: the planets, the sun and asteroids. In the 1960s and early '70s, NASA's Mariner programme sent the first human-made objects to Venus, Mars and Mercury. The Voyager programme that followed saw the first visits to the gas giants of the outer solar system: Jupiter, Saturn, Uranus and

An artist's impression of a Mars rover on the surface of the red planet.

The Mars rovers

Two extraordinarily successful robotic explorers are the NASA Mars landers *Spirit* and *Opportunity*. When they arrived at the so-called red planet in 2004, the rovers were faced with a technological hurdle: landing! The probes used parachutes to slow their descent, much like people who jump from an aircraft. Of course, the landers are a lot heavier than people, so the parachutes were very big. The parachutes slowed the landers, but they still hit the surface of Mars at around 20kph. To protect them from this jolt, the landers were fitted with airbags. Thanks to their safe landing, the rovers are still exploring the surface of Mars.

Neptune. *Voyager 1* observed volcanic activity on Jupiter's moon Io, the first active volcano discovered on another world. Today, the *Voyager 1* probe is the furthest human-made object from Earth. It is travelling out of our solar system and into interstellar space (space between the stars).

Probes continue to make important discoveries. The *Cassini* probe reached Saturn in 2004. It dropped the *Huygens* lander on to one of Saturn's moons, Titan. In 2006, *Huygens* sent back pictures of a lake made up of hydrogen and carbon on Titan. These were the first images of an extraterrestrial (beyond Earth) lake.

An artist's impression of the *Huygens* probe landing on the surface of Titan, Saturn's largest moon. The probe sent back groundbreaking images.

The use of probes has had its setbacks, however. In 2003, the European Space Agency (ESA) lander *Beagle 2* failed to contact Earth after being sent to the surface of Mars. ESA never discovered whether the lander was destroyed or made it to the surface. The Japanese asteroid probe *Hayabusa* also ran into problems. It carried a lander intended to explore a near-Earth asteroid, but a technical problem led to the lander

Robotic exploration

VIEWPOINT

Robotic explorers have achieved a great deal, and will become still more sophisticated as technology advances. Some people think they are so effective that there is no need for crewed space missions:

'We can't do much locked in a spacesuit. There isn't much to hear … The only sense that would be available to us is our eyes and we can build robots with much better eyes than humans … the little rovers on Mars right now can focus in on a distant mountain or a grain of sand. We can build telescopes on our robots with any sort of visual capability that we want.'

(Professor Robert Park, Director of the American Physical Society)

The cutting-edge *Phoenix* lander reached Mars in May 2008. It has confirmed earlier evidence pointing to the presence of water on the planet.

being released into space and floating away. However, *Hayabusa* itself was not lost and is on its way back to Earth. Should it arrive safely, the samples and data it is bringing back may prove invaluable.

✚ PROS: ROBOTIC PROBES

Probes have sent back stunning images of other planets, and have provided valuable scientific data for analysis. They have visited the outer planets of the solar system, which are too distant to be reached by human explorers with today's technology. Robotic probes have achieved this at a fraction of the cost of launching a crewed mission. This is because probes are smaller and lighter than spacecraft (which need to accommodate passengers and costly life-support systems), so they are cheaper to launch. Moreover, no human lives are at risk on these missions.

Unmanned missions into space typically take less time to prepare than crewed missions. There are no astronauts to train, and less system testing is required than when human lives depend on the reliability of the equipment. Because of these benefits probes have been used to visit all the planets of the solar system, while human crews have not been further than Earth's moon.

➖ CONS: ROBOTIC PROBES

Some enthusiasts for crewed exploration fear that the achievements of robotic probes will distract attention from efforts to achieve an increased human presence in space. Human beings are far more adaptable than any artificial intelligence that exists today. People have come up with innovative solutions to equipment failures that would have doomed unmanned missions. Some people believe a time will come when humans will want to live and work in space or settle on other planets. This will not be possible unless space agencies focus on crewed missions, develop more advanced spacecraft and find new ways to support human life in space.

CHAPTER 5

Space Technologies and Everyday Life

The space programme has helped to create many innovative technologies that improve the quality of life on Earth. Early in its history, NASA realized that the technology of space exploration had other possible applications, so in 1962 it established the Technology Utilization Program. The many space spin-offs generated over the years have had an impact on everything from medical treatments to leisure activities. These various offshoot technologies also help to promote the space programme, because they raise public awareness of its achievements and justify continued government funding.

The cool suit

Some of the most impressive space spin-off technologies are those with medical applications. One example is the 'cool suit', a system inspired by NASA spacesuit design. The spacesuits created for the *Apollo 11*

An astronaut in a spacesuit. Spacesuit technology has inspired new designs for dealing with extreme environments on Earth. Examples include thermal clothing and lightweight air supplies for firefighters entering smoke-filled environments.

lunar mission incorporated a liquid-cooled undergarment. The coolant fluid was circulated through a web of tubes by a battery-powered pump, and was intended to cool the astronauts' bodies during periods outside their spacecraft. This technique has been developed into a valuable technology for managing dangerous overheating in a variety of medical disorders.

The medical cool suit was first devised to help children who have no sweat glands. Previously, children with this condition could not take part in any physical activity, or go out on hot days. Cool suits have greatly improved the range of activities for these children. These suits also help those who suffer from neurological (brain and nervous system) problems such as cerebral palsy and multiple sclerosis. In these conditions, the body's inability to regulate its own temperature can cause heat stroke and even death. The spin-off cool suits are true lifesavers.

Microgravity medicine

VIEWPOINT

The environment of low-Earth orbit allows scientists to carry out experiments that cannot be done on the ground:

'The microgravity of spaceflight offers a unique environment for ground-breaking biotechnology and biomedical innovations and discoveries to globally advance human health in the following areas: infectious disease, immunology, cancer, ageing, bone and muscle wasting diseases, development of biopharmaceuticals [medicines made by using living organisms or biological substances], [and] tissue engineering ... '

(Dr Cheryl A. Nickerson, Professor of Life Sciences, Arizona State University)

Laser techniques

Atherosclerosis is a condition in which the build-up of fatty deposits in blood vessels restricts the circulation of blood. Eventually it may produce blockages that restrict the flow of blood to crucial organs. Blockages in major arteries can cause heart attacks, blood clots in the lungs or clots in the brain that can result in strokes.

Laser techniques developed by the space programme can now help to unclog these arteries. The technique is called laser angioplasty. It involves threading a tiny fibre-optic cable into the blood vessel. A laser beam, passing through the cable, destroys the blockage. The laser used in the treatment is called an excimer laser. In space, it is used to measure the volume of gases in the Earth's atmosphere. This laser is well-suited for its medical application because it is cool. It can be used to vaporize blockages in the arteries without harming the patient.

Other medical spin-offs

There are many other spin-off medical technologies. Digital imaging systems developed for the space programme generate pictures of faraway galaxies or the landscape of Mars. Today, they are also being used to scan patients for breast cancer. Computerized image-processing, designed to

An X-ray of a patient's chest. It shows the electronic pacemaker which regulates the heart rate. NASA technology helped to create the easily adjusted pacemakers used today.

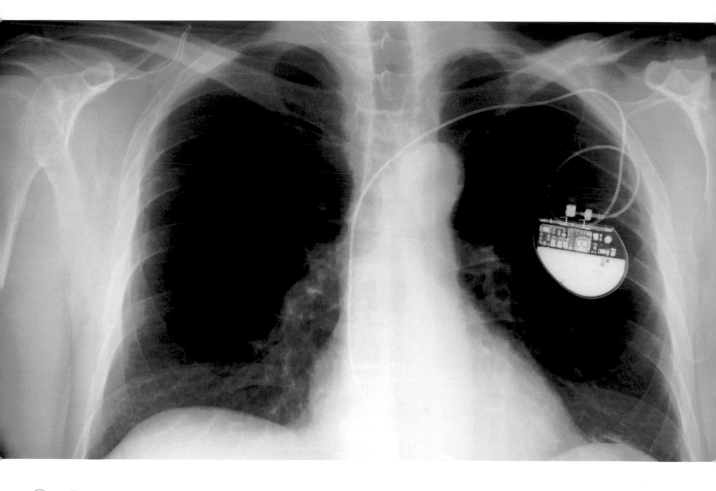

analyse the photographs taken by space telescopes, can also help to identify vision problems in children. The image-processing techniques interpret a series of pictures of the patients' eyes, examining how they respond to stimuli.

Programmable pacemakers regulate the heartbeat of patients and can be adjusted without surgery. The pacemaker uses wireless signalling technology developed to send instructions to satellites. According to NASA, one of their engineers, who was hearing-impaired, used his expertise to develop the first really effective cochlear (part of the inner-ear) implants. These implants do not amplify sound, like a traditional hearing aid, but convert it into electrical impulses. These are fed into the brain, enabling some deaf patients to hear again.

ArterioVision

Over the years, NASA has developed incredibly sophisticated software to analyse the photographs sent back by robotic space probes. NASA's programmers designed the software to pick out anything significant. Recently, this technology has been adapted for use in the medical field. ArterioVision is an offshoot of the same imaging technology, but with a very different purpose. It can create detailed computer images of a patient's arteries using an ultrasound scan. This allows doctors to detect the warning signs of heart disease and treat the condition before it becomes too serious.

✚ PROS: MEDICAL SPIN-OFFS

The merits of medical spinoffs from the space programmes are clear. They improve the life expectancies and the quality of life of people who suffer from long-term health problems. These technologies can change lives for the better.

➖ CONS: MEDICAL SPIN-OFFS

The space programme as a whole is very expensive and funded by taxpayers. Critics argue that many of the devices first developed for use by NASA would probably have been developed by scientists working in other industries anyway, and for much less money.

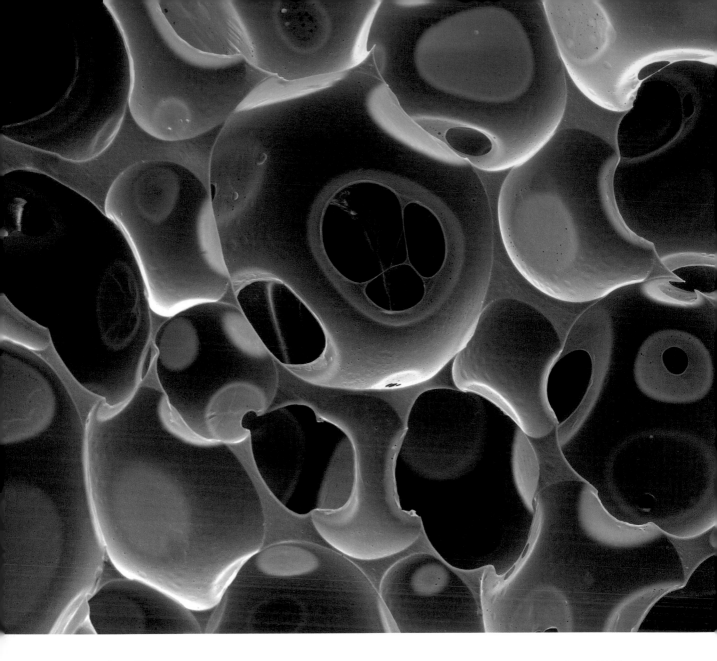

Everyday life

The science and technology of space exploration have affected everyday life in many respects, from work and travel to recreation. Space spin-offs seen in everyday life include better, brighter torches, more aerodynamic bike wheels, nutritional supplements in baby food, freeze-dried food storage, scratch-resistant lenses for glasses and more advanced cordless tools.

New synthetic materials developed for use in space have proved useful on Earth. A shock-absorbing foam, called Temper foam, was designed to withstand the pressures of high acceleration. Now it provides

Temper foam, highly magnified under a microscope. First designed for spacecraft, it is now employed for many everyday purposes.

insulation for cycling and other sports helmets. It is also used in memory-foam mattresses, which mould themselves around a person's body. These mattresses are particularly helpful for bed-bound people, preventing them from developing pressure sores.

Other synthetic materials include ceramics and plastics. Many people have braces fitted to correct the position of their teeth. The space programme helped to produce a ceramic used to create 'invisible' braces, which are less obvious than visible types. Another space-programme ceramic is added to paint to turn it into an insulator. A room painted with this special substance will retain heat better. LUNAplast, a glow-in-the-dark plastic, was also developed with help from the space programme. This material is used to make emergency signs that are visible in the dark without the need for electricity – very useful in a power cut.

Hot and cold

Technology developed for Extra-Vehicular Activity in space (see page 16) was designed to protect the body from extremes of temperature. It has proved equally helpful in dealing with very hot and very cold conditions on Earth. Smart fabric, derived from spacesuit designs, changes between a solid and a liquid in response to the temperature of the skin. It traps or disperses heat according to need. Skiers and climbers wear gear made from smart fabric to keep them warm in extreme conditions. The same material can also be used to keep firefighters cool while they battle a blaze.

Miniaturizing the computer

The earliest spacecraft did not have onboard computers – mainly because at that time computers were very large. The need to make computers small enough to fit in a spacecraft was an important step in the development of modern computing. NASA invested in the most miniaturized components available at the time – integrated circuits. The integrated circuit became a commercial success, and was soon used in other devices such as pocket calculators. By the end of the 1960s, this technology gave rise to the microprocessor and a new era of compact computing.

Air travel

Systems developed for space exploration have proved useful for commercial air travel. Techniques used for docking craft with space stations are now used to help aircraft avoid collisions. NASA technologies are also helping to reduce harmful emissions from planes, cutting down the environmental impact of air travel.

Joysticks

The technologies astronauts use to pilot the space shuttle have played a major role in the development of the modern video-game joystick controller. In the late 1990s, Thrustmaster, the company that sold joysticks to NASA, released a controller directly based on the one they had manufactured for the space shuttle. The space programme furthered the

The advanced controllers used to play modern computer games owe a great deal to space technology.

development of increasingly sophisticated and sensitive controllers for gaming. Controllers now commonly vibrate, and joysticks have a different feel or sense of resistance, depending on the fictional situation in the game. These innovations were first created for space-programme simulations, and have found their way from the space shuttle to the home.

Space technology in everyday life

VIEWPOINT

Much like NASA, ESA promotes the spin-offs from its space projects:

'... often technology developed for one application can have a previously unforeseen but highly innovative use in another. The fact that the imaging systems that we design to probe the far reaches of the universe can also be used to help uncover the innermost secrets of the human cell is indicative of the breadth of applications that can be achieved.'

(Lord Sainsbury, former Science Minister, UK)

PROS: SPACE PROGRAMME SPIN-OFFS

The technological breakthroughs that result from space exploration research often have surprising and unexpected applications. Space exploration broadens the horizons of science, and presents problems and challenges that force people to think in new ways. The results are likely to continue to improve our productivity, safety and quality of life.

CONS: SPACE PROGRAMME SPIN-OFFS

Although the space programme has produced many spin-offs, critics say that most devices developed for space missions are too specialized to be useful elsewhere. They claim that government money could be better spent on funding research and development efforts that are designed specifically to help people on Earth.

CHAPTER 6

Satellites

The satellites that orbit Earth play a silent but fundamental role in our everyday life. Satellites provide television and radio broadcasting services, phone services, weather forecasts and even driving directions. It would be difficult to imagine the modern world without them. Yet we would not have satellites without the exploration of space that began in the 1950s.

Telecommunications

Satellite technology has revolutionized telecommunications. Before satellites, phone calls travelled along cables with limited capacities. A call from Britain to the United States was transmitted by cable across the entire Atlantic Ocean! As a result, long-distance phone calls were very expensive. The first dedicated communications satellites were deployed in the 1960s, starting with *Telstar* in 1962. Today, modern satellites provide a huge number of communication channels that are used by radio, television and phone companies to send signals around the world. Satellites also transmit computer data, channelling information from the internet to users.

Orbits

Satellites can be sent into different types of orbit. Early satellites mostly occupied geostationary orbits. This means the satellite travels around Earth at the same speed as the planet spins, maintaining a fixed position in the sky. More recently, low-Earth orbits (LEO) have become common. They are cheaper because signals between Earth and the satellite cover a shorter distance and therefore require less power. In addition, LEO satellites can be serviced by astronauts from the shuttle. However, satellites of this type do not maintain a fixed position relative to the surface. So a network (or 'constellation') of LEO satellites is installed to maintain coverage.

⊕ PROS: TELECOMMUNICATIONS SATELLITES

Individuals and companies have access to a vast amount of information on the internet. Affordable international calls and email make it easy for companies to do business internationally and help people to keep in touch with family and friends.

⊖ CONS: TELECOMMUNICATIONS SATELLITES

The operational lives of communications satellites can be quite short, often around ten years. They then become space junk orbiting Earth. The main orbits are now crammed with space junk which can be a hazard to space missions and unmanned spacecraft. New satellites have extra fuel to boost them into a less populated orbit before they are deactivated. However, in 2009 two active satellites collided for the first time. The crash produced space junk that could threaten other satellites for 10,000 years. The problem is likely to become even more serious as new orbits become crowded.

A striking computer-generated image of Earth surrounded by layers of space junk. The junk includes 'dead' satellites and pieces of old spacecraft and space stations.

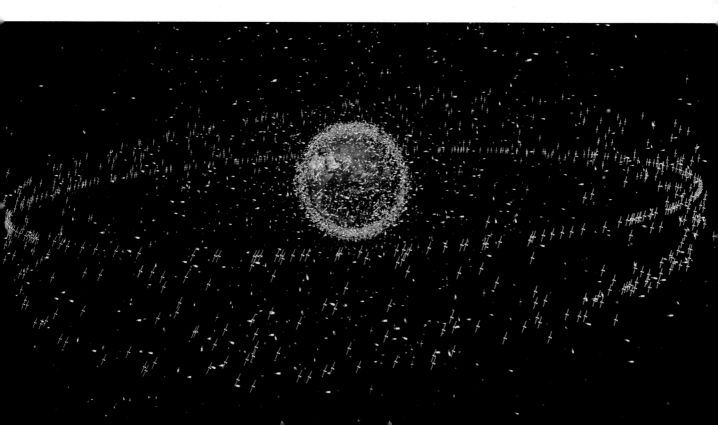

Navigation

Satellites can be used to calculate the positions of objects on Earth. This technology was originally developed for military purposes. Today it has both military and civilian applications. For example, it is used to guide vehicles and vessels, to target missiles during wars, and to help climbers and walkers locate their positions. Modern navigation satellite networks can provide coverage anywhere in the world.

The first satellite navigation system, developed by the US military in the 1970s, was the Global Positioning System (GPS). It evolved into a dual-purpose system, for both military and civilian use. The military system offers a more precise navigational reading than the civilian version, which in certain circumstances has a margin of error of up to 100m. In recent years, satellite navigation systems have become cheap

GPS systems help motorists to find their destinations. Using GPS can make journeys faster and more fuel-efficient.

enough for ordinary consumers to buy them. Modern GPS receivers are now so small that many car drivers use them as route-finders.

These gadgets are often described as GPS receivers because GPS is the dominant satellite navigation system, and the only one that is fully operational. Other networks are under construction, however. The Russians are working on GLONASS (Global Navigation and Satellite System), begun but not finished during the Cold War. When complete, the 24-satellite network will provide the Russians with global coverage. The European Union (EU) is also working on a global navigation system via ESA. The network is called Galileo, after the great astronomer.

Global Positioning System (GPS)

Navigation satellites transmit radio signals which are picked up by a GPS receiver. The time these signals take to arrive reveals the distance between the satellites and the receiver, allowing the receiver's computer to calculate its position with only a small margin of error. This calculation depends on the navigation satellites' own positions being accurately known. The orbit of each satellite should be fixed, and its location at any given time predictable. To calculate an accurate position, a GPS receiver must be able to receive transmissions from four satellites.

⊕ PROS: SATELLITE NAVIGATION SYSTEMS

Navigation satellites have increased the safety and convenience of travel. Ships are less likely to get lost in bad weather, because their GPS systems provide them with constant information about their locations. Planes are less likely to travel off course, collide or crash into obstacles. Car journeys are more efficient, as a GPS computer can plot the best route and update it if the driver goes off course or runs into problems on the road. This has a positive environmental impact, since more efficient journeys burn less fuel and reduce harmful emissions.

On a military level, GPS systems allow more precise targeting of enemy facilities. This is a great advantage for the side in control of the satellite network. Such precise targeting can reduce accidental damage to civilian structures and loss of civilian life.

CONS: SATELLITE NAVIGATION SYSTEMS

The dual military and civilian uses of navigational satellites could create problems. In the event of war, the nations that control the navigational networks might stop other countries from accessing them. This was one reason why the European Union began to develop its own system. Satellite navigation technology could also be used to destroy other satellites, since the technique that pinpoints positions on the ground can also be used to locate satellites. The possibility of 'satellite killers' destroying key enemy satellites has been a concern since the Cold War era.

Observing Earth

Several types of satellite are designed to monitor conditions on Earth from orbit. Some of the most important are weather satellites and Earth resources satellites.

The United States launched the first weather satellite, *Vanguard 2*, in 1959. Today, many weather satellites orbit our planet. Satellites allow meteorologists (weather scientists) to observe and analyse weather patterns around the world, and they have greatly improved the accuracy of weather predictions.

Satellite imagery

Earth observation satellites collect information by recording electromagnetic wavelengths. They do this by measuring the radiation naturally reflected by the Earth or by directing an artificial beam at the planet and observing the reflected energy. They store this data and then transmit it to Earth for analysis. Computers can use the data to produce digital images which contain a great deal of information. These satellite images are analysed and compared using sophisticated software. Depending on what is being studied, these images can reveal crucial information, such as the size and location of oil spills or changes in the levels of the various gases in the atmosphere.

It is vital to form a picture of the weather over the entire planet, because conditions in different regions influence one another. Before the satellite era, it was impossible to forecast accurately more than a day ahead. With the help of modern computing, the data from weather satellites enables forecasters to predict conditions over several days.

Knowledge of weather conditions helps to make people safer. Weather information aids navigation at sea or in the air. Advance warning of extreme weather events can give authorities enough time to plan emergency measures or to evacuate threatened areas.

A satellite picture of a typhoon over the Pacific Ocean. With the help of weather satellites, governments can plan for such dangerous events.

Earth resources

Earth resources satellites are used to observe and learn about conditions on Earth. These low-orbiting satellites provide data used in geology (the study of the Earth), cartography (mapping the world), oceanography (the study of Earth's oceans), forestry (studying forests and monitoring deforestation) and volcanology (the study of volcanoes). For example, observation satellites can use infrared radiation to measure changes in soil composition and detect mineral deposits. They can map changes to coastlines caused by changing sea levels, identify the spread of oil spills and track the gas given off by volcanic eruptions.

Many satellites focus on studying the Earth's plant life, in particular crops, with networks such as the long-running French SPOT (*Satellite Pour l'Observation de la Terre*, meaning Satellite for Earth Observation). NASA's ongoing Earth Observing System (EOS) programme seeks to monitor changes in global conditions over the long term.

Earth observation satellites provide invaluable data about the Earth's resources and ecosystems. They also allow scientists to monitor humanity's impact on the planet. For example, data from a satellite, *Nimbus 7*, confirmed that there was a hole in the ozone layer over Antarctica. This hole is an area of damage to Earth's protective atmosphere that allows harmful radiation from the sun to pass through. International action has reduced the emissions that caused the damage, and this has allowed the ozone layer to begin to repair itself.

Earth observation satellites are used to monitor the activities of people as well as conditions on the planet. Spy satellites (also called reconnaissance satellites) are a special type of Earth observation satellite, put into orbit by governments for political or military reasons. Spy satellites

The importance of satellites

VIEWPOINT

There is no doubt that the ability to launch satellites has transformed daily life:

'Whether most people realize it or not, outer space is now part of everyday life for much of the planet's population. Satellites have become fundamental to modern society ... Broadcast television, the internet, ATM machines, banking transfers, telephone service, credit card validation, weather prediction, terrestrial and oceanic mapping, atmospheric and natural disaster monitoring, urban planning, navigation, and even targeting of sophisticated weaponry all rely on the use of satellites.'

(Theresa Hitchens, Director of the Center for Defense Information, United States)

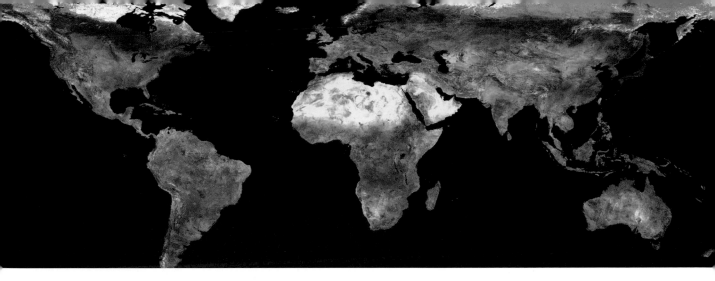

are designed to gather intelligence about the activities of actual or possible rival states. Scientists developed satellite technology at the height of the Cold War rivalry between the United States and the USSR, so spy satellites were rapidly built and have been orbiting ever since. The data they provide includes information about the military strength of rival states. They can also be useful in detecting signs, such as troop movements, that suggest a state may be planning to launch an attack.

This world vegetation map was created using information from several Earth observation satellites. Deserts and other dry areas range from yellow to brown, while forests are green. Snow and ice are blue and white.

➕ PROS: SPY SATELLITES

A state that possesses spy satellites will feel more secure. It knows what its enemies are doing and can respond in good time. Spy satellites could be seen as helping to preserve peace, since an aggressive state is less likely to prepare for war, let alone launch an attack, if it knows that its would-be victim will be ready for it. In a situation where rival states both have spy satellites, they are less likely to take military action against each other.

➖ CONS: SPY SATELLITES

In the real world, states are seldom on an equal footing. If a great power has aggressive intentions, advanced spy satellites will increase its advantage. The information it possesses about its rivals may actually encourage it to go to war. Like most forms of technology, spy satellites are only beneficial if used responsibly.

CHAPTER 7

The Economics of Space

Economics is the study of how goods and services are produced, bought and sold, distributed and consumed. Economists study the way resources are used. Consumers must decide what purchases are worth spending their money on. When governments and private companies think about investing in space exploration, they have to make the same kind of decisions. They have to decide whether the benefits of exploring the universe justify the cost. Studying the economics of space involves trying to assess these costs and rewards.

Paying for the space programme

Sending people and machines into space is very expensive. Space programmes require large numbers of people and considerable amounts of equipment. Europe's ESA employs over 2,000 people and has an annual budget of around €3 billion. This is less than a quarter of the size of NASA's budget, which was US$17.3 billion in 2008. These costs have led some people to question whether investing in space exploration provides value for money.

Economic criticisms   **VIEWPOINT**

The cost of the space programme has been a focus for criticism since its earliest days:

'And I say to you today, that if our nation can spend US$35 billion a year to fight an unjust, evil war in Vietnam, and US$20 billion to put a man on the Moon, it can spend billions of dollars to put God's children on their own two feet right here on Earth.'

(Martin Luther King Jr., US civil rights leader)

PROS: SPENDING ON SPACE

Although the costs are high, supporters of space exploration argue that its benefits easily justify the expense. The information about the origins and nature of our universe provided by telescopes, probes and crewed missions is of great scientific interest. There are not always immediate material or financial rewards from this knowledge, but supporters argue that expanding humanity's understanding of how the universe works is worthwhile in itself. Today, some space activities pay for themselves. Private companies pay state-run space programmes to launch their satellites. Then they get their money back by selling their satellites' services – such as telecommunications or television – to consumers. Corporations also pay to have experiments conducted in microgravity conditions.

US astronaut Sunita Williams at a training centre outside Moscow, Russia. The men and women who work at the world's space agencies receive expensive and specialized training.

CONS: SPENDING ON SPACE

Space exploration is expensive. Critics of the space programme argue that money would be better spent elsewhere. Humanity already possesses enough space technology to service our everyday needs (primarily to launch satellites into orbit). Trips further into space, back to the Moon or to Mars, are unlikely to produce immediate material returns.

Technologies developed for space are frequently useful on Earth. However, opponents believe it is better to try to solve Earth's problems directly, rather than adapting technologies created for space. They also argue that the immediate needs of people on Earth outweigh the importance of learning about the universe or exploring other planets in our solar system.

Private enterprise and space

For decades, businesses have been hired by state-funded space programmes to manufacture parts and provide expertise. However, private companies have only recently developed their own spacecraft. In the future, space may become accessible to many more individuals through private services.

In 1996, the X Prize Foundation announced a competition to build a reusable private spacecraft. The US$10-million prize would go to the first private spaceship to travel into space and return to Earth twice in a two-week period. The challenge inspired the company Scaled Composites to begin work on a vessel, with financial backing from

SpaceShipOne, the first private craft to travel into space, takes off. Instead of rising on the back of a rocket, the craft is lifted into the air by a special plane.

SpaceShipOne

SpaceShipOne was designed by the aerospace engineer Burt Rutan. The craft was built to be light so that it could be carried by an aeroplane and launched in the air. This was possible because it did not need to achieve full orbital velocity. However, the team behind SpaceShipOne is now hoping to develop an orbital craft. If so, it may overcome the limitations of contemporary spacecraft such as the Russian *Soyuz*, the Chinese *Shenzhou* and the US shuttle, which all rely on disposable elements.

the co-founder of Microsoft, Paul Allen. In 2004, their spacecraft, named SpaceShipOne, completed the first successful private crewed expedition into space.

The spacecraft briefly left Earth's atmosphere, reaching an altitude of 100km. That's not high enough to go into orbit, but SpaceShipOne's trip was a landmark in the history of private space flight. A second flight a few days later secured the US$10 million prize. The same team are currently working on improved versions of their ground-breaking craft.

➕ PROS: PRIVATE SPACE TRAVEL

The involvement of private firms may speed up innovation. Many of the craft of the national space programmes have been in service for a long time. New designs are limited by restricted budgets. Russian *Soyuz* spacecraft have been in service since 1967, and are still used to take astronauts to the ISS. NASA's space shuttle has been around since 1981, and many people believe it is time for a new vehicle.

Private firms want to maximize profits and minimize costs. This may be a spur to producing a fully reusable craft. Private sector contributions could help public space programmes, and research and development by private companies may compensate for the limited government funding available for developing new ideas.

➖ CONS: PRIVATE SPACE TRAVEL

Not everyone is in favour of private companies having an independent involvement in space. Some fear that increasing private activity (sometimes called the commercialization of space) will shift the focus of space efforts away from scientific exploration to profit-making gimmicks. Space activities are generally governed by international treaties, many of them drawn up before any question of private space travel arose. There is debate over what laws would be needed to regulate the activities of private companies in space and who should make those laws.

Space tourism

The first space tourists paid the Russian national space agency to take them into orbit. The Russians were the first to offer this service, in conjunction with the company Space Adventures, because they urgently needed to raise funds during the difficult period following the collapse of the USSR in 1991.

American billionaire Dennis Tito on board the International Space Station in 2001. Tito was the first private individual to pay to be taken into space.

The first paying space traveller was Dennis Tito, an American businessman with a scientific background. He paid a reported US$20 million to be taken up to the ISS, where he carried out his own experiments. Since Tito's trip, others have followed his example and paid to be transported to the ISS aboard Russian *Soyuz* spacecraft.

Now, however, the next phase of space tourism seems about to begin. The Virgin Group is preparing a service called Virgin Galactic. It will regularly carry space tourists on sub-orbital space flights, where they will briefly experience weightlessness and enjoy spectacular views of the Earth from space. The craft are based on a modified version of SpaceShipOne.

SpaceShipTwo *Enterprise* is the first of the new ships. It was set to begin test flights in 2009. Although the price of a trip into space was US$200,000 for the first 100 tickets, they have already been booked. Virgin aims eventually to bring the cost down to a level that more people can afford. If large numbers of people take up space travel, there may be a demand for extended holidays in space. This might lead to the opening of space hotels, a concept already being developed by several companies.

The benefits of private space travel

VIEWPOINT

In a video interview with the *Daily Telegraph*, Virgin Group chairman Sir Richard Branson enthused about private space flight and Virgin Galactic's future plans:

'I think [space tourism] will be the beginning of real space exploration. I think that once you get private companies in there … we'll be able to let people experience the joys of space. We'll be able to send satellites into space at a fraction of the costs that have been set in the past. Once our engineers have finished this programme, they'll straight away be moving on to seeing if we can fly passengers between London and Australia in around about half an hour, just popping the space ships out of the Earth's atmosphere and back down again …'

(Sir Richard Branson, chairman, Virgin Galactic)

➕ PROS: SPACE TOURISM

Space tourism could allow private individuals to explore space. Astronauts have spoken of the emotional impact of seeing the Earth from space. In the future, many more people may share their experiences. A thriving space tourism industry could become an active part of the world economy, creating large numbers of jobs and using the products of other industries. The aim of any business is to make money, so a space tourism industry will have an interest in devising the cheapest way to carry people into space. This could produce innovations which will be of great significance to any future human voyages into space.

➖ CONS: SPACE TOURISM

Space flight consumes valuable natural resources, both in the production of the craft and in fuelling each launch. Each trip to space involves burning rocket fuel, which gives off pollutants. These include chlorofluorocarbons (CFCs), which damage the ozone layer. Most people agree that current space flight is responsible for only a small portion of CFC emissions. But if space tourism becomes a booming industry, these pollutants could pose a serious problem.

Space tourists put their health in danger. Professional astronauts undergo a training that lasts far longer than the three days offered by Virgin Galactic. Astronauts are extremely fit, carefully selected individuals. Although passengers will have to pass physical tests to qualify for private flights, it is difficult to predict how well tourists will be able to withstand the stresses of launch and microgravity.

There are also concerns about the fitness of the spacecraft. Private companies are profit-driven and aim to cut costs. National space programmes can afford to invest more in safety, yet there have still been tragic accidents. Critics fear that the odds of a disaster on a private space flight will be much higher. It is up to the future tourists to decide if the risk is too high.

The dangers of private spaceflight

VIEWPOINT

The actor William Shatner played Captain James T. Kirk of the Starship *Enterprise* in the TV science-fiction programme *Star Trek*. Shatner turned down a free seat on the first trip of Virgin Galactic. He summed up fears over the safety of private space travel, saying:

"… to vomit in space is not my idea of a good time. Neither is a fiery crash with the vomit hovering over me … "

Chinese tourists take part in a brief exercise designed to familiarize them with microgravity. Training like this will be a standard aspect of future space tourism.

CHAPTER 8

Our Future in Space

So far, humans have travelled only into nearby space, sending satellites and space stations into orbit, and setting foot on the Moon. Space agencies and space exploration enthusiasts hope that we will venture farther out. New technologies may revolutionize our ability to travel into and work in space. Some of these breakthroughs may come soon, while others will remain confined to sci-fi stories. However, writers described many inventions before the technological know-how existed to build them. Pioneering science-fiction writer Jules Verne imagined travelling to the Moon in *From the Earth to the Moon* (1865) a century before the Apollo landing. Today's fantasy may be tomorrow's reality.

New technologies for new frontiers

Space exploration benefits from advances in a number of scientific fields. Nanotechnology, for example, represents a new stage in the trend towards miniaturization. It uses very small materials or devices, measuring 100 nanometres (millionths of a millimetre) or less. The most cutting-edge nanotechnology involves manipulating matter on a molecular level, assembling materials atom by atom. Nanotechnology will probably lead to the manufacture of materials that are stronger and lighter than anything available now. This could make space travel cheaper, since a lighter craft requires less fuel to launch. Nanotechnologies will also allow the creation of tiny computers and machines. One day there may be miniscule microprobes capable of exploring the universe.

New technologies may also overcome one of the greatest limitations of current spacefaring: the speed of spacecraft. When *Apollo 11* went to the Moon, it took over four days to travel the 384,400km. At present, a return trip to Mars would take up to three years. A truly revolutionary propulsion system might even unlock the galaxy beyond our own, allowing humans to cover the distances between solar systems, which are measured in light years. Research programmes, such as NASA's Breakthrough Propulsion Physics (BPP), investigate whether there is any hope of travelling faster than the speed of light. In the meantime, humans have plenty to explore in their own solar system.

An artist's impression of a space elevator made out of a futuristic material. In the future, technology like this may allow us to reach orbit at a fraction of present-day costs.

The space elevator

The space elevator is a futuristic concept that could change the way that objects are put into orbit. The idea is to use a long tether to link an orbiting station with a similar station on the ground. A lift would go up and down the tether between the stations, transporting its cargo into orbit for a fraction of the cost of a rocket or shuttle launch. As yet, no material exists that is strong enough to make the tether. However, nanotechnology is already producing thinner, tougher materials. Further advances could make the space elevator a reality in the not-too-distant future.

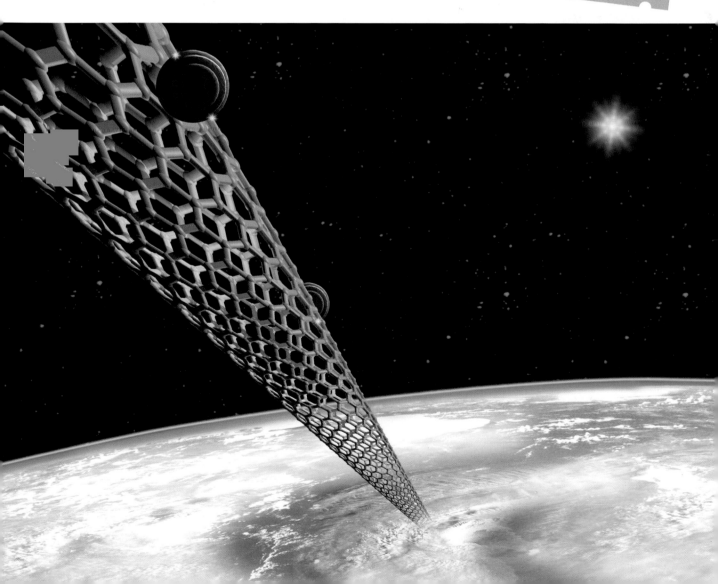

> **➕ PROS:** NEW TECHNOLOGIES
>
> New technologies may transform the role of space exploration in human life. Humans have mastered the use of Earth's near space for satellites, but have not settled in space or exploited the resources of other worlds. However, innovative technologies may open up the final frontier, perhaps allowing humans to become true spacefarers.

> **➖ CONS:** NEW TECHNOLOGIES
>
> New technologies are likely to give individuals or countries a kind of power – for good or for evil – that has never been seen before. Humanity has been highly destructive as well as creative, so the prospect is not an entirely comfortable one.

The search for life beyond Earth

Humans have long been fascinated by the possibility of life on other planets. Some scientists believe that they have found traces of very basic life forms in meteorites from Mars, and that microscopic life may once have existed there. However, not everyone is convinced. Scientists still disagree on how likely it is that there is life – and especially intelligent life – anywhere in the universe besides planet Earth.

In the twentieth century, there were several attempts to detect signs of alien intelligence. Most focused their efforts on scanning for extraterrestrial radio waves. Radio astronomers are still on the lookout with their radio telescopes, searching for transmissions from other stars. Their efforts are complicated by the many frequencies available and by sources of interference such as cosmic background radiation and interstellar gas.

The SETI (Search for Extraterrestrial Intelligence) project evolved in the 1960s from attempts to detect alien radio signals. The scientists used radio telescopes, such as the enormous one at Arecibo in Puerto Rico. There were some moments of great excitement when the search appeared to have paid off, but these seem to have all been false leads.

The Arecibo radio telescope in Puerto Rico is the largest telescope on Earth, measuring 305m across. For the past half century, radio telescopes have scanned outer space for alien signals.

One transmission that was believed to be from Epsilon Eridani turned out to have come from a passing aeroplane. After a brief period of enthusiasm, policy makers became increasingly reluctant to use public money to search for extraterrestrial beings.

The search for alien intelligence has yet to yield any concrete results. But the universe is very big! Many enthusiasts remain determined to continue the search for life somewhere else in the universe.

The SETI League

Today, private sources provide most of the funding for the search for alien life. Anyone who is interested can get involved. The SETI League is a non-profit organization. It supplies free software for download, which members can install on their computers. The program allows thousands of household computers to help process vast amounts of radio data. Using many ordinary computers to analyse data in this way is a technique called distributed processing. It is a cheap way of processing data without having to buy very powerful specialized computers.

➕ PROS: THE SEARCH FOR ALIEN INTELLIGENCE

Some people are inspired by the possibility that other intelligent life exists in the universe. The discovery of extraterrestrials could transform our world. An exchange of ideas and technologies with an advanced alien race could help the people of Earth in ways that cannot even be imagined.

➖ CONS: THE SEARCH FOR ALIEN INTELLIGENCE

Searching for extraterrestrials takes time and money. Many experts think it is highly unlikely that there is alien intelligence on other worlds. Others believe that, while alien life may exist, it is probably too far from Earth for astronomers to hope to detect any sign of it. Even if humanity did make contact with an alien civilization, its members might not be friendly.

Extraterrestrial resources

The solar system contains many natural resources that we cannot access. In his 2004 Vision for Space Exploration, US president George W. Bush re-committed the United States to the goal of establishing a permanent lunar base. The Moon is known to be rich in iron, silicon, aluminium, titanium and tritium. The first thorough harvesting of extraterrestrial natural resources is likely to begin there. But in the current state of technology, transporting raw materials between the Moon and Earth would be very expensive. As a result, most of the resources mined would probably be used in space.

One exception is energy. Because of the absence of an atmosphere in space, solar panels constructed there

VIEWPOINT

Space resources

For years, space pioneers have looked forward to the day when humanity will be able to make use of the resources of the cosmos:

'Future generations will say that the real significance in our space program lay in the fact that it took the lid off the limitations posed by the finite size and finite resources of the planet Earth ...'

(Werner von Braun, scientist and designer of the V-2 rocket)

could harvest the sun's energy very efficiently. The energy could be cost-effectively transported, via a microwave link, to a receiver on Earth.

Somewhat further in the future, nations or private companies may set up mining bases on asteroids. The solar system contains a large number of asteroids, which are filled with valuable mineral deposits. Iron asteroids could be particularly rewarding, as they contain tons of iron, nickel, platinum and gold. It is likely that early efforts will focus on the asteroids close to Earth, but once spacefaring technologies are more advanced, prospectors may work on the vast asteroid belt that lies beyond Mars.

This artist's impression shows an astronaut mining on an asteroid. Mineral resources mined in space could prove to be a valuable by-product of space exploration.

⊕ PROS: SPACE RESOURCES

Many of the Earth's natural resources are non-renewable. Once they are all used up, they are gone forever. Mining resources from other bodies in the solar system could remedy shortages on Earth. The demand for metals such as the iron found in asteroids is always increasing, as are humanity's energy requirements. If new space industries develop, they are likely to assist economic growth and employment. Finally, new resources with unique applications may be discovered on other worlds.

⊖ CONS: SPACE RESOURCES

It is by no means certain that it will ever be practicable to tap extraterrestrial resources. If it does happen, such resources could provide an easy fix for shortages caused by our wasteful patterns of consumption. Humanity might come to depend on these new resources, instead of devising more economical, sustainable and environmentally-friendly practices.

Colonizing space

Many science-fiction stories imagine a future in which humanity has moved out into space and settled on other worlds. This may happen one day. But at present, the technology needed to create self-sufficient cities on other planets does not exist. It will probably be a long time before scientists and engineers overcome all of the problems involved.

However, humankind is already taking the first steps towards the permanent settlement of space. Space habitats are housing teams for increasingly long stays. The ISS is the most sophisticated space station built to date, and its modular design leaves room for it to expand to accommodate larger teams in the future. NASA's plan to build a Moon base within 20 years may prove even more significant. Eventually, perhaps, it will lead to the creation of an outpost on Mars. Enthusiasts for the colonization of space hope that these developments will prove to be the first steps towards large-scale human settlement of the galaxy.

Terraforming

One day scientists may be able to alter the ecology of inhospitable planets to make them suitable for human habitation. This process is called terraforming. Various methods of terraforming have been suggested and imagined. They all involve changing a planet's climate and atmosphere to make it more like Earth so it could support plant and animal life. Mars is the most frequently discussed candidate for terraforming. In his Mars trilogy, American writer Kim Stanley Robinson vividly imagines the transformation of Mars from a red planet to a planet of oceans, flora and fauna.

➕ PROS: SPACE COLONIES

Some people believe that colonizing space is the only way to ensure the long-term survival of the human race, since human civilization on Earth will eventually come to an end. At the latest, this will happen when the sun exhausts its supply of hydrogen fuel, in an estimated five billion years. Of course, some other unforeseen disaster could wipe out all life on Earth before then. For example, a devastating asteroid impact might occur, similar to the one that made the dinosaurs extinct.

⊖ CONS: SPACE COLONIES

The colonization of space represents a great departure from human beings' natural habitat. Some people oppose this on a moral or religious basis, believing humans are not intended to leave their world. The effects of living on other planets could lead the settlers there to evolve in new directions, and to develop unique cultures, a prospect that some find troubling or even threatening. An alternative view is that colonization on any scale will never be practicable. If so, plans based on the idea will prove to be expensive fantasies.

An imaginary picture of colonists at work on Mars. The idea of colonization has excited people for a long time. But Mars is is cold and airless – very different from Earth. Colonists would have to live in enclosed buildings and vehicles.

GLOSSARY

artificial intelligence The computer-based ability of machines such as space probes to calculate and make decisions.

asteroids Rocks in space, mostly orbiting in a belt between Mars and Jupiter. Though some measure up to 1,000km in diameter, they are too small to be regarded as planets.

Big Bang The event that caused the universe to begin to expand, and created all of space and time.

ceramic A non-metallic material created by heating one or more inorganic substances. Common ceramics include glass and pottery, but new, high-tech ceramics are used to create many things, from heat-resistant fabrics to artificial bone implants.

constellation (1) A group of stars which, when viewed from Earth, appear to be close to one another, often forming a distinctive pattern. (2) A group of satellites that work together to perform a function such as mobile phone coverage or GPS.

dwarf planet A rounded, planet-like body that orbits a star, but does not qualify as a planet. The most famous dwarf planet is Pluto, which was considered the ninth planet of our solar system until 2006.

electromagnetic spectrum The range of possible wavelengths of electromagnetic radiation that an object can give out. The visible light we see with our eyes is just one small part of the electromagnetic spectrum.

European Space Agency (ESA) A cooperative European effort to explore space. It was founded in 1975, and now includes 18 member nations.

extraterrestrial Originating outside our planet Earth.

Extra-Vehicular Activity (EVA) Any activity that involves astronauts leaving their spacecraft.

galaxy A group of stars bound together by gravity.

gas giant A large planet that is made up mostly of gas, rather than solid matter such as rock. The outer four planets of our solar system – Jupiter, Saturn, Uranus and Neptune – are all gas giants.

geostationary orbit An orbit in which the object travels around Earth at the same speed as the planet spins on its axis, and so remains above the same point on the planet's surface.

GLOSSARY

gravity The force that pulls objects towards one another. The greater the mass of an object, the more powerful its gravitational pull.

insulation A material used to prevent the transfer of heat, in order to keep something hot or cold.

integrated circuit An electronic circuit that has been manufactured on to a single circuit board, rather than constructed out of separate components. The integrated circuit was crucial to the development of miniaturized electronics. It is also known as a microchip.

laser A device that fires a beam of light, not always from the visible portion of the electromagnetic spectrum.

light year The distance that light travels in one year, 9.46 trillion kilometres.

low-Earth orbit (LEO) An orbit in which the object is no more than 2,000km in altitude. To date, all space stations have occupied a low-Earth orbit.

matter The physical substance that things are made of. Anything that takes up space is made of matter, whether in solid, liquid or gaseous form.

microgravity The condition commonly described as 'weightlessness', in which the Earth's gravitational pull is counteracted by the speed of an orbiting spaceship or space station.

microprocessor An integrated circuit containing the central processing unit (CPU), or brain, of a computer.

module An independent component made to a fixed standard. This allows it to be connected to and interchangeable with other compatible modules.

nanotechnology Technology that operates at a microscopic level, involving units measuring 100-millionths of a millimetre or less.

National Aeronautics and Space Administration (NASA) The United States' space agency, responsible for the federal government's programme of space exploration.

orbit The path an object follows around a larger object as a result of its gravitational pull.

ozone A gas in the upper atmosphere that prevents harmful ultraviolet radiation from reaching the surface of Earth.

GLOSSARY

planet A body that orbits a star. A planet has enough mass to have become rounded by the force of its own gravity. It is larger than an asteroid, but not large enough to generate energy through nuclear fusion (as a star does).

pulsar A small but extremely dense collapsed star that spins and sends radiation out into space.

quasar The super-hot core of a faraway galaxy that releases enormous amounts of energy.

satellite An object that orbits a planet. The Moon is Earth's only natural satellite.

solar system All the bodies that are subject to the gravitational pull of the sun.

space probe An unmanned mechanical device, usually fitted with a range of scientific equipment, launched into space.

star A ball of hot gas, held together by the gravity of its own mass, that releases large amounts of energy.

sub-orbital Describes an object that reaches space but returns or falls to Earth without going into orbit.

synthetic Manufactured by humans. Synthetic substances do not exist in nature, but are created by combining different materials or by altering an existing material in some way.

telecommunications The long-distance transmission of audio and video communications signals.

ultrasound High-frequency sound, beyond the range of human hearing. Ultrasound can be used as a non-harmful means of scanning for medical and industrial purposes.

vacuum The absence of matter and atmospheric pressure.

velocity An object's speed and direction.

FURTHER INFORMATION

WEBSITES

http://www.nasa.gov/externalflash/nasacity/landing.htm
The NASA Home and City application provides an interactive tour of some space spin-off technologies.

http://hubblesite.org/
Website for the Hubble Space Telescope, including a gallery of spectacular images.

http://www.space.com/
Online community for space news and information.

http://www.virgingalactic.com/
Virgin Galactic's website, promoting their planned space tourism service.

BOOKS

Dorling Kindersley Visual Encyclopedia of Space
Professor David W Hughes and Robin Kerrod,
Dorling Kindersley (2006)

The Cosmos: A Beginner's Guide
Adam Hart-Davis and Paul Bader,
BBC Books (2007)

Space Guides: Discovering the Solar System
Peter Grego, QED Publishing (2007)

The Inside and Out Guide to Spacecraft
Clare Hibbert, Heinemann (2006)

Navigators: Technology
Peter Kent, Kingfisher (2009)

Is There Other Life in the Universe?
Kate Shuster, Heinemann (2008)

Impact of Science and Technology: Communications
Andrew Solway, Franklin Watts (2009)

INDEX

Page numbers in **BOLD** refer to illustrations and charts.

air travel 34, 39
alien life 54–5
Andromeda Galaxy 5, **5**
angioplasty 30
Apollo 11 8, 10, 28, 52
Apollo-Soyuz missions 14
Armstrong, Neil 10
asteroids 4, 24, 26, 57, **57**, 58

Big Bang 7
Braun, Werner von 8, 56
Bush, George W. 12, 15, 56

Cold War 8, 11, 12, 14, 39, 40, 43
computers 30, 33, 36, 39, 40, 52, 54
cool suits 29
Copernicus, Nicolaus 6, 7

electromagnetic spectrum 22–3
European Space Agency (ESA) 22, 26, 35, 39, 44
Explorer 1 9
Extra-Vehicular Activity (EVA) 16, 33

Gagarin, Yuri 9, 16
galaxies 4, 5, **5**, 22, **23**, 24, 30, 52, 58
Galilei, Galileo 6, 7, 39
Global Positioning System (GPS) **38**, 38–9

Hawking, Stephen 7
Hubble Space Telescope 22, 23, **23**
Huygens probe 25, **25**

international cooperation 14–15, 56
International Space Station (ISS) **14**, 15, 19, **19**, 20, 47, **48**, 49, 58

joysticks **34**, 34–5
Jupiter 4, 6, 24

Keck Observatory, Hawaii 22
Kennedy, John F. 10

lasers 11, **13**, 29–30
light years 5, 53
Luna 1 24

manufacturing in space 21
Mars 4, 24, **24**, 26, **26**, 30, 45, 52, 54, 57, 58, **59**

Mars rovers 24, **24**
medical technologies 29–31
memory foam **32**, 32–3
Mercury 4, 24
microgravity *see* weightlessness
Mir 14, 16, 18, 19
Moon 6, 8, 10, **10**, 24, 27, 44, 45, 52, 56, 58
  base 56, 58
  exploration 10, **10**, 24
  landing 10, **10**
  resources 56

nanotechnology 52, 53
National Aeronautics and Space Administration (NASA) 8, 9, 10, 16, 18, 22, 24, 28, 31, 33, 42, 44, 47, 53, 58
  funding 31, 44, 45
  space spin-offs 28–35
  Technology Utilization Programme 28
Neptune 4, 25

ozone hole 42, 50

pacemakers **30**, 31
Polyakov, Valeriy 16
probes 22, 24–7, **25**, 31, 45
pulsars 7

quasars 7, 23

Reagan, Ronald 11
robots 22, 24, 26, 27, 31
rockets 8, 16, 17, 50, 53
Rutan, Burt 46

Salyut 18
satellites 8–9, **9**, 11, 12, **13**, 18, 31, 36–43, 45, 49, 52, 54
  collisions 37
  communications 36–7, 45
  Earth observation 40–43, **43**
  military use 12, **13**, 42–3
  navigation 38–40, 41
  orbits 36
  weather 40–41, **41**
Saturn 4, 24, 25
Saturn V 8, 16–17, **17**
Search for Extraterrestrial Intelligence (SETI) 54, 55

Skylab 19
solar system 4, 6, **6**, 24, 27, 46, 53, 56
space elevator 53, **53**
space junk 37, **37**
space mining 56, 57, **57**
space programmes 10, 11
  cost 11, 44–6
  private enterprise 46–7, 49, 51
  spin-offs 28–35
space race 8–11
space resources 15, 56–7, **57**
space settlement 58–9, **59**
space shuttle 14, 16–18, **17**, **20**, 23, 34, 35, 36, 46, 47
space spin-offs 28–35
space stations 14, **14**, 15, 16, 18–21, **19**, **20**, **48**, 52, 58
space technologies 28–35, 46
space tourism **48**, 48–51, **51**
SpaceShipOne **46**, 46–7, 49
SpaceShipTwo *Enterprise* 49
spacesuits 16, **28**, 28–9
Sputnik 1 8, 9, **9**, 10, 11
Star Trek 4, 51
'Star Wars' 11
stars 4, 5, 6, 7, 22, 24, 25, 54
Strategic Defense Initiative (SDI) 11, 12
sun 4, 5, 6, 7, 14, 24, 42, 57, 58

telecommunications 36–7, 45
telescopes 6, 22–3, **23**, 30, 45, 54–5, **55**
Tereshkova, Valentina 9
terraforming 58
Tito, Dennis **48**, 49

universe 4–5, 6, 22, 23, 44, 45, 46, 52, 54, 55
  size 4–5
  understanding 6–7, 23
Uranus 4, 24

Venus 4, 24
Verne, Jules 52
Virgin Galactic 49, 50, 51
Voyager 1 24–5

weapons in space 11–13, **13**
weightlessness (microgravity) 19–21, **20**, 29, 49, **51**